WINGS No.5

Marine Muscle:
Hornet a...

Hans Halberstadt

Windrow & Greene

© 1993 Windrow & Greene Ltd.
Published in Great Britain 1993 by
Windrow & Greene Ltd.
5 Gerrard Street
London W1V 7LJ

A CIP catalogue record for this
book is available from the British
Library.

ISBN 1-872004-47-4

Published in the USA by
Specialty Press Publishers
& Wholesalers Inc.
PO Box 338
Stillwater, MN 55082
(612) 430-2210/800-888-9653

Dedication:
For Jay Francis Ensor

Acknowledgements:
Harry Truman once said that the Marines had
a propaganda machine the equal of Stalin's,
and he was at least partly right. The US
Marine Corps is unique among the five
American armed forces for managing to
inspire a kind of devotion and affection quite
unapproached elsewhere. Marines are nearly
always evangelists for the Corps. They love to
talk about it - its history, its mystique, its
mission, and its many characters. Producing
books about Marines is a treat because the
Marine Corps is a great story.

I owe a salute to the Headquarters Public
Affairs shop and the ADCO, who helped
support the project and got me a Hornet ride.
The PAO shops at Cherry Point, North
Carolina, at Yuma, Arizona, and at El Toro,
California, provided really superb support.
VMFA(AW)-225, with Lt.Col. Jon 'Goose'
Gallinetti presiding, were among the most
gracious and pleasant of all the units I've ever
worked with. Everything about this squadron
sparkles. Thanks especially to Major Lance
Olson and Capt.'Tinker' Bell for the baby-
sitting service, the interviews, and particularly
the ride. A pat on the back also to Major Bill
'Mace' Macak, Capt.Tom 'Nuts' Donahoe and
Capt.Todd 'Fewts' Kemper. Capt.Dave 'Blade'
Bonner, formerly of VMA-231 (now of
American Airlines) was another of those
Marine evangelists, and spent a lot of time
briefing me on the Harrier II, its operations in
the Gulf, and Marine aviation in general.

Thanks to all these, and the other Marines
that helped out. As they like to say, *'Semper Fi!'*
Hans Halberstadt
San Jose, California
October 1992

**(Title page) An AV-8B Harrier II of
squadron VMA-311 'Tomcats' on the
flight deck of the USS *Belleau Wood*
during Operation Tandem Thrust.**

Marine Air

Cruising along at about 18,000 ft, a Hornet flight from Marine squadron VMFA(AW)-225 'Vikings' enjoy the sights en route to the Nevada training ranges. North of Los Angeles is a vast, desolate area of desert and arid mountains where fighters are permitted to frolic and spar without generating too many noise complaints. In the distance is California's highest mountain, Mt.Whitney (4,418 metres). Although the terrain below is sparsely populated, it is a notoriously rotten place to practise forced landings or parachute descents, so if you're going to have an emergency, have it someplace else.

The US Marine Corps occupies a unique place in the American military community because of its very special mission and legendary status. The mission is to be ready to put a credible combat force across any beach in the world, and keep it there, no matter what. To do that the Corps has built a force and a doctrine that is among the very best anywhere. To the Marines the Corps is something more than just another military community; to most, it is a kind of calling, inspiring feelings not unlike those found in some religious communities - somewhat like Jesuits with rifles.

It is a community with a mystic sense of self that is entirely different from any other American armed force. Its people are held to a far higher standard, its missions are among the most challenging, and generally Marines claim a willingness to offer a great deal more suffering and sacrifice than the other services. There is an extreme sense of loyalty and commitment to each other, and to whatever their mission may be. Traditionally this is expressed in two words: *Semper Fidelis* - 'always faithful' - signifying that a Marine would rather die than fail. And that means that Marine Air is

Somewhere high over the Pacific Ocean Lt.Col. Jon 'Goose' Gallinetti, commander of VMFA(AW)-225, and his faithful WSO, Executive Officer Major William 'Mace' Macak, cruise above the clouds. Lt.Col. Gallinetti's aircraft is designated '01', and is the official squadron Hornet. It has been embellished with an artificial canopy on its underside to bewilder the opposition (an innovation of those fiendishly clever Canadians, adopted by only a few American operators), and sports a pair of AIM-9 Sidewinders and a belly tank. The FLIR pod is visible inboard - one of the key systems that make this D-model such a superb attack aircraft, offering all-weather targeting and navigation.

dedicated - first, last, and always - to supporting Marines on the ground, at whatever cost.

That's where the Hornet and Harrier II fit in the Marines' mission, as essential components of the Air Combat Element, in direct support of the Ground Combat Element. One way or another, when Harrier IIs or Hornets fly they are adding muscle to the Marine rifle companies manoeuvring inland from the beach. Both aircraft are extremely popular within the Marine Corps for their *Semper Fi* commitment. The Hornet gets high marks for its marvellous ability to do many things well; it has been called the finest fighter in the world, and is a superb interdiction bomber. The V/STOL Harrier II is the world's premier close air support aircraft, able to put bombs on target within ten minutes of a forward air controller's call. With the Harrier II flying close air support and the Hornet providing both combat air patrol and battlefield interdiction, the Marine Corps has a lot of muscle to flex in combat. Both these aircraft were risky decisions when they were selected, but have turned out to be immensely successful in their chosen roles.

The Hornet is certainly not the fastest jet in the sky, but it isn't supposed to be. Agility and effective targeting/engagement of threats in the air have become more important than raw speed; and the Delta version of the Hornet, with its dedicated weapons systems officer, provides the Marine Corps with a platform that can fly both combat air patrol (CAP) and close air support (CAS) on the same mission. A laser spot tracker is attached to the starboard intake. The fuselage hump contains most of the aircraft's fuel, about 1,400 gallons, with some 200 in the wing tanks.

1: Sting of the Hornet

If you only look at the specifications, the F/A-18 might seem inferior to a lot of other combat aircraft in service today. The MiG-29 and Su-27 are both far faster in the horizontal plane; the F-16 has a better thrust-to-weight performance, and can fly straight up and away from a fight, while the Hornet (particularly the D-model) lacks power in certain flight regimes, particularly at low-level; and even the old A-6 Intruder that it replaces has a better range.

But despite all that, the Hornet is considered by many - both within and outside the US naval aviation community - to be the best fighter around. Why? Because of its combination of agility and weapons systems, and the way the pilot is integrated with the machine. Other aircraft have great virtues and areas of individual superiority, but the Hornet has a combination of talents that can help it dominate the skies in a way that Eagles and Falcons often cannot.

It is good at air-to-air and ground attack; other aeroplanes can do one or the other well, but the Hornet is world-class at both. In the words of one long-time Intruder pilot, new to the Hornet: 'The difference between the Hornet and the A-6 that I used to fly is night-and-day. The A-6 was a great aeroplane: you could carry a lot of bombs, and you had "long legs"; the F/A-18 doesn't have as long legs and doesn't carry as much ordnance - but it has extreme accuracy.

'I have just about the least amount of time in the Hornet of any pilot in the squadron, but I can consistently put all my bombs within 30 feet of the target with the F/A-18. With the A-6, a 100-foot CEP (Circular Error Probability) is considered pretty good. We can't carry as much ordnance, but we can put it a lot closer. In our business, that's what it's all about.'

The Marines have always been notorious for pinching pennies, but they all agree that the Hornet has been a bargain - two aeroplanes for the price of one (well, one and a half, with cost overruns). It is essentially the same jet that the US Navy uses, but the Corps have a different mission for it

Although the WSO has a long list of cockpit chores to attend to, and no flight controls to get in the way, he still has to help keep an eye out for enemy aircraft. Despite the appearance of spacious accommodations, the cockpit is a tight squeeze that you don't exactly get into but rather put on.

Here's the view from the No.2 aircraft in a three-ship trail formation, with one ahead and one astern (check six, starboard).

concentrating primarily on supporting the Marine infantry and armour attack across the beach. One Hornet crewman explained it in the following terms:

'The Navy and Marines both use the Hornet. For the Navy, the aircraft supports the ship primarily in the air-to-air and power projection roles; for the Marine Corps the primary mission is close air support - supporting the Marine on the ground. That includes defending a strike package from enemy fighters as well as responding to that FAC (forward air controller) on the ground who's calling in close air support.

'Although the Harrier II has the ability to engage airborne targets, it does not have a radar; the F/A-18 is the fighter, for all intents and purposes, in the Marine community. During the Gulf War the Hornets were able to drop bombs on their targets, come off, and then shoot down enemy aircraft if required to do so.'

For pilots coming out of the attack community, the broader mission of the Hornet is a new

challenge. These Marines are used to putting bombs and bullets on target. They still do that, with improved accuracy and ease of delivery; but they have to learn to be skilled air-to-air fighter pilots as well.

'I'm starting to learn about the air-to-air mission; it is *really* difficult. The amount of time you need for planning and briefing is tremendous. Coming from the A-6 attack community I find that this part of the mission is really challenging. The attack mission is so *easy* in this aeroplane - the Hornet does it for you - but the air-to-air mission is a different story!'

Flying the Hornet: Preflight

Well, it's time to get suited up and go try one out. Collect your gear from the Flight Equipment shop, and we'll go out to the flight line.

The D-model Hornets of VMFA(AW)-225 are all lined up, clean and smart. The 'Vikings' squadron have new aircraft which still look sleek, elegant and fresh out of the box. The precise latitude and longitude for each parking space is stencilled on

Although it's not really polite to line up on the skipper's aircraft, everybody does it anyway. He's a nice guy (unless you shoot, of course), and will probably do the same for you. This is a rather unusual shot in that the Marine Corps have previously been reluctant to allow photography of powered-up cockpit displays. The left CRT (Cathode Ray Tube) shows the HUD symbology, including a lock-on, while the NAV data is in the centre, and the weapon system solution (here, guns) is on the right.

(Right) Lined up for MCAS El Toro's main runway pair, Seven Left and Right, and trailing vapour from the soggy coastal air, Lt.Col. Gallinetti leads 'Goose' flight back to the deck at 145 knots, sliding through 3,990 ft toward Seven Left.

the concrete; that's for the inertial navigation system calibration that allows you to take off, fly for hundreds of miles, and still know almost exactly where you are.

A walk-around inspection reveals a really beautiful jet only about half the bulk of the F-15 or F-14, with the fit and finish of the best road cars. Start at the boarding ladder and work your way around clockwise, following the preflight checklist. A peek inside the access panels reveals that all the black boxes are latched and connected right where they belong.

All the panels are secured, leaving the skin of the aircraft unbroken. The leading edge of the right wing reveals the drooping leading edge slats that extend and retract under computer control. The wing tip is equipped with a missile launch rail, and there is an AIM-9 Sidewinder installed. Trailing edge flaps are fully extended - huge 'barn doors'

that let the fighter both slide back to earth at a reasonable airspeed and get airborne in a hurry.

At the stern, look into the engine exhausts for anything loose, burnt or missing. These General Electric F404-GE-400 turbofan engines are superbly responsive and reliable. Tucked between the nozzles is the tailhook for those occasions when the aircraft operates from one of the Navy's carriers, or needs to make an emergency landing back at base.

Climb under the aeroplane, still looking for anything worn, loose, disconnected, broken or missing. Well, the landing gear is still there, looking like cast iron - this gear is made for carrier landings, the most abuse you can give a set of tyres. But these are in perfect shape, so continue the inspection out along the wing. Inspect the weapons stations; all the bombs and missiles are secured, the safety wires installed properly, without

any play. The armourers have loaded the jet with a full combat load: medium range AIM-7 Sparrows on the engine nacelles, short range AIM-9s on the wing tips, Mk 83 bombs on the wing pylons, and shells for the 20mm cannon. Instead of the Mk 83s they could have hooked on Mavericks, HARMs, napalm or Paveways - soon, AMRAAMs will also be added to the Hornet's arsenal.

As far back as the 1960s there was a school of thought that considered the cannon an obsolete weapon for a fighter, but Vietnam War experience indicated otherwise. The M61A1 20mm version fitted in the Hornet will fire about 4,000 rounds per minute (6,000 in extremis). The whole gun and ammunition package can be removed and replaced as a single assembly; but all you see of it before flight are the apertures in the nose for the shells, and the gas deflectors.

Preferred weapons for most missions are the 500-, 1,000- and 2,000-pound free-fall bombs, occasionally with the Snakeye fins installed. The weapons are normally mounted on the VER-2 vertical ejector rack, which pops the weapons off the aircraft with greater precision than a simple unlatching mechanism. Besides the Maverick, which can be either laser- or infrared-guided, the Hornet drops the Walleye (eletro-optical/TV tracker), the Rockeye (anti-tank cluster bomb), the

HARM (a broad band anti-radar missile), and the Harpoon (anti-ship missile). While many aircraft can deliver these weapons, the Hornet is designed to make the delivery highly effective and easier than with previous systems through its digital databus and computer combination. But you can't see much of that walking around the aeroplane.

Finally, after a close look at everything, and having listened to the assurances of the crew chief that the Hornet is airworthy, you inspect the book for any logged deficiencies and sign on the proper line. Now it's all yours.

Lighting the fire

You board by climbing a little ladder on the port side, an ingenious device that folds down out of the wing root. Have a look down into the aft cockpit; it's tight, the seat is firm, and the view is dominated by three television-type screens. There are no flight controls back here, but there is plenty to keep the Weapons Systems Officer (WSO) occupied.

Like most contemporary fighters, the Hornet is about as easy to start as your car - but you don't need a key. Once you've climbed aboard and strapped in, find the power panel over by your right knee; switch the battery power switch forward to ON. Then locate the auxiliary power unit (APU)

Short final. Most Hornet recoveries
are carrier approaches, the most
difficult and challenging for Marine
pilots - and the most dramatic and
theatrical for observers, since the
technique involves slamming the
aeroplane back on the deck, in
contrast to the gentle, smooth
landings of land-based aircraft.
Unlike carrier decks, however, the
runway at El Toro only rocks and
rolls during earthquakes.

switch on the left console and move it to ON; a little gas turbine engine will start up automatically, providing power to the aircraft. Once it gets up to speed (usually after about 20 seconds) the light for the APU within the cockpit will glow green. The right engine gets started first: if there is a fire, and if it isn't doused by the self-contained fire extinguisher mounted in the engine bay, you still have the option of retreating down the aircraft ladder on the left rather than ejecting.

Move the right engine start switch to START. Monitor the panel for fuel flow, temperature and RPM, and when it gets to 15 per cent RPM slide the right throttle forward from STOP to IDLE. Listen to the turbine sing its rising song; the RPM indicator will start to twitch, then climb. The engine lights off with a gentle *whoosh,* and the RPM will begin to soar. The aeroplane awakens from its slumber at about 64 per cent RPM; the hydraulic pressure comes up quickly and the electrical generator comes on line. The left engine starts the same way. Within minutes the aircraft is almost ready for flight.

The inertial navigation system needs time to think - about six minutes for accurate initializing. Then call the tower: *'El Toro Ground, Viking 41, flight of four with information "Bravo", taxy to marshal.'* The clearance comes back almost instantly, and the flight is cleared to the marshalling area just down the taxyway to wait for everybody to get in line. When all four aeroplanes are 'good to go', the flight leader calls the tower again: *'El Toro Ground, Viking 41, flight of four with information "Bravo", taxy for take-off.'*

The Hornet taxies like a BMW - fast, with a firm, flat-footed suspension that is a delight. Pre-take-off checks get done during the taxy-out. And as with most squadrons, everything is done with precision, even down to the closing of canopies: *'Canopies down',* Lead calls on the radio, and all four lower together. As the canopy comes down the outside world becomes suddenly quiet. The inside world is quiet too, except when the tower calls: *'Viking 41, position and hold.'*

Taxy to the edge of the runway and hold short, while a pair of single-seat C-models slide down the glide path for the other runways. With the gear and flaps out in the breeze and the power throttled back, the Hornet settles out of the sky like a controlled brick, slamming aboard in good carrier form. When Flight Lead gets a 'thumbs up' from the rest of the flight, he will call the tower: *'Tower, Viking 41, flight of four, take-off, IFR.'*

'Viking 41', the voice replies, *'cleared for take-off.'* Immediately the flight changes frequencies to departure control, and calls: *'Departure, Viking 41, deck check, cleared to one three thousand.'* With toe brakes set, run the throttles up to military power (about 80 per cent) for a last engine check. No worries? Normal procedure is a 'herd go', a multiple

or formation take-off. *'Release brakes - NOW',* Flight Lead calls. Release the brakes, throttle forward into 'burner' - and off you go.

Acceleration is modest for about 2.75 seconds; then the afterburner kicks in. The jolt is a firm, authoritative push like that of a muscular car - but it doesn't stop. Within seconds you are roaring down the runway at about 140 knots indicated air speed; bring the stick smoothly aft, just enough to lift the nose wheel from the deck. The wheels will promptly become unstuck.

With a positive rate of climb, and only when Lead calls *'Gear - NOW',* select the gear to UP, and the flaps will automatically retract when the flight control computer tells them to. *'Burner - NOW'* is

Lead's next call. Bring the throttle back to MIL power: you sag forward in the straps as the acceleration eases from 'Warp Nine' to a modest 'Seven' or so, which happens to provide about 300 knots airspeed - normal climb-out velocity under 10,000 feet.

That airspeed will usually yield about 4,000 to 5,000 feet-per-minute in a steady climb. Within a minute of becoming airborne you are steaming along through 5,000 ft AGL (Above Ground Level) in the clouds. The tops are at 6,000 ft, and after another few seconds of climbing through the soup you come banging out into brilliant sun and an empty blue sky. And that is basically how you get a Hornet into the air.

The Hatching of the Hornet

The F/A-18 is certainly one of the finest combat aircraft available anywhere today - according to one Air Force evaluator, the best overall fighter in anybody's skies - but it is in fact the direct descendant of a design that came off second-best in a competition 'shoot-out'. The victor was the prototype of today's F-16, and the 'shoot-out' took the form of a fly-off, held in 1975, to decide which prototype best matched the requirements set by the US Air Force for their 'air combat fighter' contract. That particular dogfight used statistics, computer models, test pilot evaluations and huge quantities of politics to decide the issue - and the Falcon is certainly a winner in anybody's book.

The AN/AVQ-28 HUD fitted in the Hornet cockpit is not quite as big as that in the F-15, but it is still a wide-field, bright, full-service display that incorporates a director gunsight. This photo has been slightly 'sanitized', but still shows the heading (254 degrees), angle of attack, altitude and air speed. 'I used to laugh at people when they told me they could see me at ten miles - but with the HUD display helping, you can actually see somebody at fifteen.... He is inside this little box on the HUD, and once you can actually see him it makes engaging him much easier. It is so easy to be flying at low level and still be within five seconds of your time-on-target in this aircraft, thanks to the display on the HUD'.

(Right) The front seat panel is elegantly simple in the D-model, particularly compared to earlier generations of fighter aircraft with their 'steam gauge' instrumentation. The three CRTs can be programmed to show almost anything except the afternoon 'soaps'. In this case we have a status display on the left, showing the condition of many aircraft components. The centre display shows the moving map, complete with current position; this is driven by the extremely accurate inertial navigation system (INS). On the right is the built-in-test display.

But the US Navy needed a new aircraft to serve alongside its new, very expensive and very large F-14 Tomcat; and the YF-17 (as the prototype was known) caught the attention of the selection board with its promise of low cost, high agility, twin engines, small size, and potential for development.

The first Hornet flew off the deck on 18 November 1975, and it was love at first flight. The jet was both nimble and quick. There were the usual teething troubles, but the aeroplane was demonstrating all kinds of qualities that were overlooked in the Air Force competition. It was so good that the Navy decided that separate versions for fighter and ground attack missions weren't needed - the same airframe could accomplish both

tasks. And while the costs and complexities of the project increased, the Hornet team took wry satisfaction from the news coming from the F-16 shop, where the engine that was supposed to be so reliable wasn't, and the costs that were supposed to be so low were heading through the roof.

Although there was some scepticism and apprehension about the aeroplane at the beginning, the pilots were soon raving about it. During trials the F/A-18 was 'hosing' the opposition, no matter what they were flying. Even when it took on its bigger brother, the F-14 Tomcat, the Hornet came out on top: in one series of 34 engagements the Tomcat failed to score against the Hornet even once. And according to a report in the Spring 1990 issue of *World Air Power Journal,* the Hornet had a chance to spar with a MiG-23 over the Nevada test ranges and 'cleaned the clock' of its Soviet-designed adversary. The Navy and Marines duly ordered 1,157 examples.

When the first airframe was handed over to Fleet Replacement Squadron VFA-125 'Rough Raiders' at Lemoore, California, in May 1980 it had 'Marines' painted on one side and 'Navy' on the other. The first real operational unit was a Marine squadron, VMFA-314 'Black Knights' at Marine Corps Air Station El Toro, who opened for business with the Hornet in January 1983.

Marines use aeroplanes a little differently than the other services do. The Air Force and Navy fighter communities lean heavily toward the air superiority mission, but the traditional mission for Marine pilots is down in the mud with the rifle companies, putting a 'world of hurt' on the opposition. As one pilot put it: 'Marine Air has one mission, and one mission only - to support the infantryman. How we do that depends on tactical circumstances. It can be electronic warfare, anti-air, air defence or offensive air support - all aid the Marine on the ground.

'It's the Marine Ground Combat Element (GCE) commander who determines how that air will be used. His colleague, the Air Combat Element commander, knows how to employ air in a given situation; it might be by sending missiles up tailpipes to maintain air superiority, insuring Marines aren't getting bombed or strafed. It can be by offensive air support - hitting targets that effect the Marine Expeditionary Force. It can be battlefield interdiction - hitting targets that might affect the battle in the future, or taking out a bridge to keep enemy reinforcements from reaching the battle.'

This tradition actually goes back to the 1920s, when a small expeditionary force in a remote corner of Nicaragua got into a fight with some of the locals in the middle the night. A patrol of slow old DH-4 biplanes arrived overhead, figured out what was going on, and started firing their Great War surplus machine guns. When they ran out of ammunition they went back to collect some bombs, and returned to the party. The 25-pound 'pig iron' devices were delivered from a shallow dive, very close to the ground; and close air support was invented! It worked like a charm, too: the Marines on the ground discovered that the volume of fire directed at them slackened, then stopped. The survivors evaporated into the jungle, leaving about 50 of their *compadres* in various states of disrepair outside the Marine position. Since then the CAS mission has been the first priority for Marine aviators.

*   *   *

It is hard to believe that the design is 25 years old, but it's true. The Hornet was created at about the same time that absolute speed became less of a factor in fighter specifications than agility, range,

(Left) The cockpit seems almost spartan by comparison with many other fighters, but that's actually an advantage. The idea is to let the pilot keep his attention outside the cockpit as much as possible, and to make it easy to select weapons, designate targets, change radio frequencies, drop bombs and perform all the other chores without ever taking his hands off the throttle and stick - the HOTAS concept ('hands on throttle and stick'). The seat is the Martin-Baker US10S (SJU-5/6), with zero-zero capability.

(Right) The weapons stores can all be discarded in an emergency with the simple push of a button, without actually shooting them.

The engine panel, on the lower left side of the instrument panel, uses liquid crystal displays to provide information in a simple, graphic way. The 'Bingo' gauge indicates the minimum level of fuel required to return safely to base.

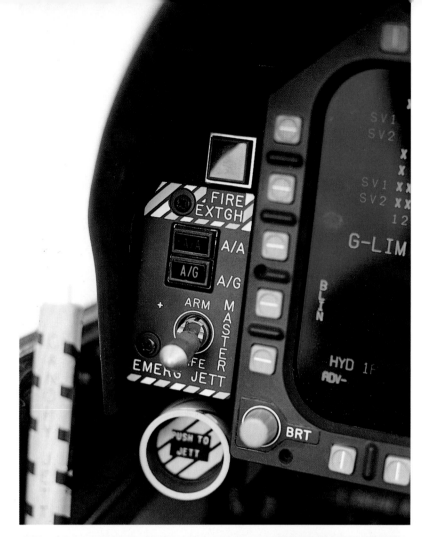

radar performance and weapons integration. Until then fighters like the MiG-25 (a Mach 3 performer) and the F-4 expected to zoom rapidly upward to high altitude, intercept a flight of intruding bombers or fighters, engage with radar- or IR-missiles, then go home. But battle tactics changed, and the bombers went low; so it was back to the drawing board for fighter designers around the world.

If the Hornet came out of the nest looking a lot like the MiG-29, Su-27, F-14 or F-15 it is because all use similar specifications for a similar mission. There are two fundamentals to these specs: the physical size of the radar antenna, and the turn performance required in engagements. The former determines the size of the nose, the latter the shape of the fuselage.

All these aircraft use twin engines for survivability, but the twin engines mandate a shape that suffers from a bad case of midriff bulge. That causes problems for controllability at slow speeds, because tail surfaces get masked by air intakes unless countermeasures are taken. The solution to the problem is the same for Russian and American designers alike - twin tails, vortex generators, and leading edge extensions.

(Above) Over by your right knee are some of the housekeeping controls: radar altimeter, hydraulic pressure gauge, generator switches and the battery switch, along with a reminder check list for landing.

(Left) That 'castle'-shaped button is one of the keys to the HOTAS system; you can use it to change weapon displays and select ordnance that has been programmed into the computers. The trim switch is on the right. Each has a distinctive shape and feel to make identification easy even with 'eyes out' of the cockpit.

(Right) Inside your left knee is the panel for the countermeasures systems. Although the aircraft can automatically dispense flares and chaff to spoof IR missiles, and electrons to spoof acquisition systems, the pilot or WSO can also do so manually. (And according to the display, we're fresh out of everything.)

Specifications and dimensions

The Hornet, as delivered, turned out to be about half the size of the Tomcat: a petite 37,000 lbs in the single-seat fighter configuration, spread over a fuselage 56 ft in length, and a wing span of 37 ft 6 in diameter. Inside the slender fuselage were a pair of F404-GE-400 turbofan engines, each of which could put out 16,000 lbs of thrust. The engines are identical, which makes exchange a lot easier; and each puts out enough thrust to get the airframe home should the other engine quit. Even so, the Hornet in its D-model configuration is hardly the powerhouse of the fighter community. You can point the nose at the sky, apply afterburner and pray, but forget any illusions of zooming vertically toward outer space in this jet.

'If you pull the NATOPs regs out they will tell you that the aeroplane has a one-to-one thrust-to-weight ratio; but if you put a couple of Sidewinders on there, and maybe a drop tank, no way! In my old A-6 you could just about fly around with the canopy open, but drag really affects this jet. Now, the Hornet is a superior fighter, but it is a *slow* fighter. You can't run away from a fight with an aeroplane with the dash capability of, for example, the F-16, because it will just run you down.'

Without any weapons on the wings the Hornet could wind up to about Mach 1.8 - not nearly as fast as a MiG-29's Mach 2.5, but still pretty quick.

And when you load the Hornet up, there is room for four air-to-air plus five air-to-ground weapons. You can hang just about anything on the racks of anybody's fighters - MiGs and Hornets, and nearly all the other fighters across the globe, use the same 14-inch lug spacing for ordnance, a legacy of World War 2. But using a weapon effectively is another matter; and that has a lot to do with the target acquisition, designation and guidance systems that are part of the aeroplane, rather than the weapon. The Hornet's systems permit the use of all the 'dumb iron' bombs up to the Mk 84 2,000-pounder, but almost anything in the inventory with pylons can do that. It's the synthesis of computer, HUD, and pilot training that makes the Hornet valuable, because it puts the bombs right where they will do the most damage with greater precision than earlier systems.

Then there are the precision guided munitions, the 'smart' weapons. These need terminal guidance, either through the use of lasers, infrared imagery, or by manual 'steering' by the WSO controlling a video image. Smart bombs need smart aeroplanes and crews to be effective, and the Hornet is designed to be among the brightest. The maximum warload works out at about 9,200 lbs of bombs, missiles and shells: 18 Mk 82 500-pounders, for example, or six Mk 83 1,000-pound bombs, plus a pair of Sidewinders, would be typical cargoes.

The twin F404-GE-400 engines toe out at the intakes a bit, are perfectly identical, and each produce about 16,000 lbs of thrust on a good day. Although the specs say that works out to a one-to-one thrust-to-weight ratio, don't believe it. The Hornet will go straight up about as well as it will reach Mach 1.8 - empty, unloaded, with the squadron's smallest pilot up front. But raw speed is not the measure of this airplane - agility is. The engines respond quickly and reliably, giving the Hornet the kind of manoeuvrability that is still, after all these years, a crucial element of air combat. The big arrestor gear is stowed between the exhausts.

(Right, above) The intakes are separated considerably more than the nozzles, a function of the off-centreline alignment of the engines. Since the aircraft is designed for speeds well below Mach 2 the design foregoes variable geometry intakes for an elegant D-shaped form with a fixed splitter plate. The AN/AAS-38 FLIR pod gets latched into the recess on the port side intake.

(Right, below) If the markings on these aircraft were any more subdued they would be invisible. The circular inlets are cooling air spill louvres, and the stubby blade is a small 'fence' that helps spread the vortices generated by the wing's leading edge extension. The fence is particularly valuable when the aircraft is at high angles of attack, as the tails normally loose aerodynamic effect, resulting in the rudders losing authority.

Don't expect your squadron mates to be particularly reverential, no matter what your rank or score. This major's name is just too close to that of a popular brand of American dog food ('Alpo') for his compadres to ignore - even if he did put some serious metal on the Iraqis in the Gulf War. Most of these mission markings are gone from the aircraft now, and the rest are fast disappearing.

(Right, above) Port side fin tip embellishments, top to bottom: AN/ALQ-165 high-band ECM transmitting antenna, AN/ALQ-67 radar attack warning antenna, another AN/ALQ-165 antenna, and the fuel dump outlet. The red navigation light contains a strobe.

(Right, below) The squadron commander's aircraft markings are a little sharper than everybody else's, but they are still a far cry from the gaudy, high-visibility paint schemes of the past. Some squadrons dispense with virtually all ornamentation, but VMA(AW)-225 retains a 'Viking' warrior on all its airframes.

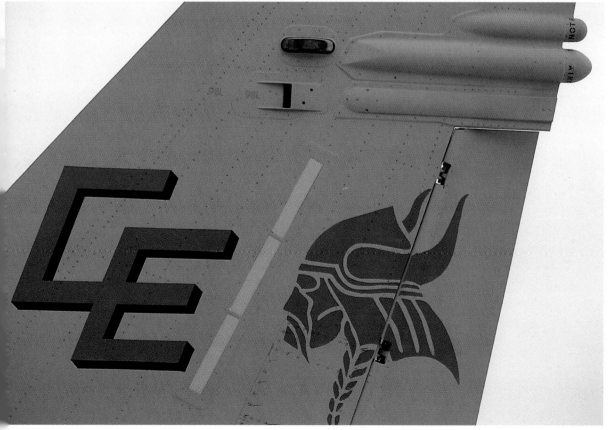

VMFAT-101 'Sharpshooters' are the training squadron at MCAS El Toro.

(Right) The squadron commander's aircraft has rather unsubdued markings, including the chequerboard pattern used by the 'Death Rattlers' for decades. This is an A-model aircraft with a lot more hours on it than the 'Vikings' D-models, and the wear is beginning to show.

(Below) VMFA-314 took the honours for first operational use of the Hornet, beating the Navy for the prize. The 'Black Knights' participated in the attack on Libya in 1986, attacking surface-to-air missile sites during *Operation Prairie Fire*. Perhaps as a result, the markings on this Hornet are hardly subdued at all.

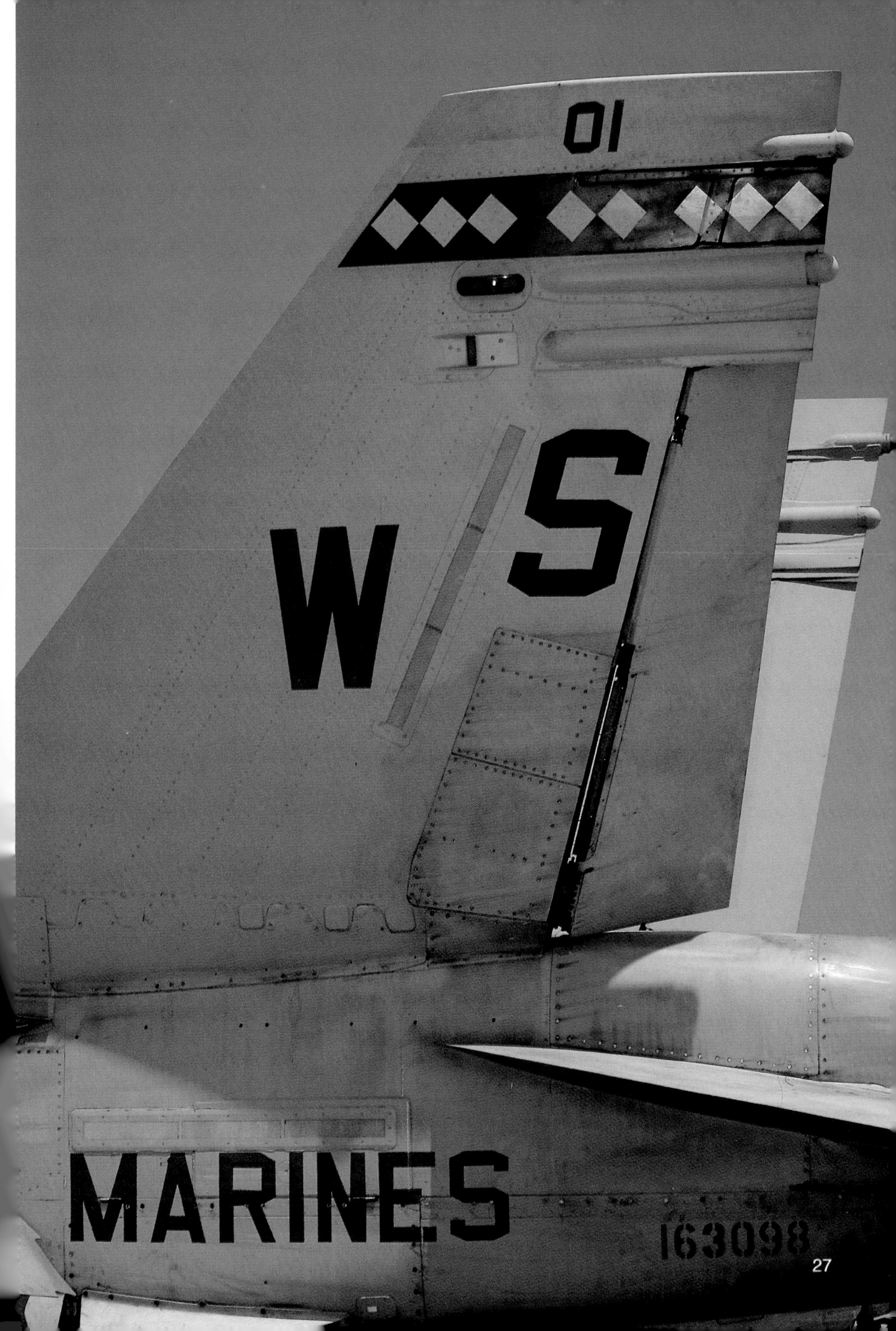
01
W S
MARINES
163098

One of the many enlisted 'Vikings' inspects the variable nozzle of a D-model Hornet. Everybody in the squadron, pilots included, must wear boot covers when atop the aircraft, which may have something to do with the pristine condition of the paint jobs.

A paint shop Marine adds a fresh 'Viking' to a newly acquired Hornet.

(Right, above) Squadrons find ways to individualize their aircraft, and paint jobs are only superficially uniform. This airframe's mini-fence has the manufacturer's hornet logo applied to it. The owner/operator is VMFA-323, nicknamed the 'Death Rattlers'.

(Right, below) The FLIR pod is about to have its innards inspected. It may not look like much, but this equipment is worth considerably more than its weight in gold, especially when there is 'pig iron' that needs to be delivered close to a Marine rifle company in heavy contact with the enemy. Then the pilot can see and engage targets on the ground, through darkness and rain, with a high degree of accuracy.

Tail assembly detail on an AIM-9 Sidewinder. The blue colour indicates an inert rocket motor and warhead.

(Right) The little bump houses one of the ALR-67(V) radar attack warning (RAW) antennas; the pop-up indicators are safety devices.

(Right) A pair of AGM-88 HARMs (High-speed Anti-Radiation Missiles) wait in the ordnance compound for the call, after which they will have fins attached. The HARM is a marvellous missile that homes on radar energy of almost any tactical frequency. That means you can take out (or frighten out of business) the operators of the many kinds of radar that are used to track and engage 'strike packages' - and you can do it from long range, with a very high probability of a kill. The HARM is one of the most successful weapons in the arsenal, and another reason for the success of the Allied air campaign during the Gulf War.

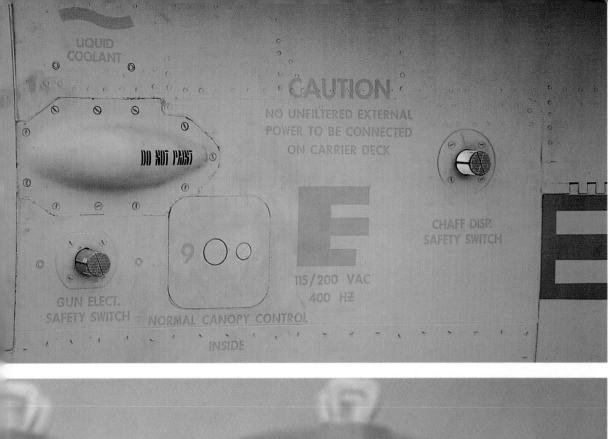

LIQUID
COOLANT

DO NOT PAINT

CAUTION
NO UNFILTERED EXTERNAL
POWER TO BE CONNECTED
ON CARRIER DECK

E

115/200 VAC
400 HZ

CHAFF DISP.
SAFETY SWITCH

9 ○ ○

GUN ELECT.
SAFETY SWITCH

NORMAL CANOPY CONTROL

INSIDE

An F/A-18A from VMFAT-101 slides back toward the numbers and good old Zero Seven Left at El Toro.

Preflight inspection rituals are the same for almost any American combat aircraft; you start at the cockpit and work your way around in a clockwise sequence.

An F/A-18D Hornet in the nest.

(Below) With the exterior checks complete it is time to climb in and continue with the cockpit checks. The Hornet's boarding ladder is built right into the port LERX, and folds neatly out of the way when not needed.

Ground crew tidy up before launch, closing the canopy access panel and preparing for engine start.

Proper kit for Marine aviators includes a lot of emergency equipment, including this floatation device for water survival. The beads on the activation handles provide a touch identification for a flyer who may not be able to see them; a pull on the beaded handle will inflate the device and pop him to the surface.

Standard items in the vest include a signal mirror, 'pencil' flare gun and flares, a smoke marker, water bottle and whistle.

(Right) The oxygen mask is a crucial piece of equipment for Hornet crews, and the positive pressure regulator makes it more comfortable to use. Although the cockpits are pressurized, the system that keeps them that way can sometimes fail without the pilot noticing until it is too late. The result is called hypoxia, and it kills aviators with surprising frequency.

Once all the checks are complete the crews wait for everybody else in the flight to get squared away. If one of the aircraft has a problem, that can be a long wait.

'Starting Two', signals
Lt.Col.Gallinetti.

The Australian Hornet driver patch is extremely good trade bait in Marine units.

Despite a tragic inability to speak American, Australian F/A-18 pilots like Flt.Lt.Paul Proudfoot participate in an exchange programme that provides familiarization with Allied tactics, procedures and innovations. Somewhere 'down under' is a lonely US Marine serving in an Aussie squadron, where his new mates can't understand him either.

Capt.'Tinker' Bell (left) and Major Lance Olson, two of VMFA(AW)-225's team of Hornet drivers. Capt.Bell is a convert to the Marines, having previously spent a good bit of time in the US Army's 82nd Airborne Division. Major Olson is the squadron's high-powered operations officer.

VMFA-242's version of the unit insignia.

2: Swoop of the Harrier

An AV-8B Harrier II owned and
operated by VMA-211 'Wake Island
Avengers' at Yuma, Arizona. This
night attack variant is loaded up
with little practice bombs and is all
fired up and ready to go. However,
the rest of the flight is still getting
its act together.

The US Marines claim a lot of heritage from their British cousins. Many of the Corps traditions come from an attempt to pattern the US Marine Corps along the lines of the Royal Marines after the American Civil War, and all new Marines learn that their origins were in the 'fighting tops' of English men-of-war prior to the American Revolution. The Harrier II is another British legacy that the Marines have adopted with enthusiasm.

In fact, it is one of the success stories for the Corps, a weapon that happens to fit the Marine mission perfectly. That mission is always focused on the Marine amphibious assault, the projection of power by sea to bases ashore. And the Harrier II helps that mission along better than any aircraft the Marines have ever owned.

The Harrier II, as used by today's US Marines, is the third or fourth generation of the breed. The aircraft was born as the Hawker Siddeley Kestrel in 1961, taking flight in a form quite recognizable today. It was tested in 1961 by the RAF, US Army, Navy and Air Force and the German Luftwaffe; and initially rejected by all but the RAF. (British development continued, of course; and the world

Before a new Harrier II driver gets to go and play with the Fleet he must first get a ticket-punch here at Yuma's dusty carrier qualification pad. The seas are always calm here, even when the winds pick up - which is just as well for pilots new to the Harrier, an aeroplane that lands on a pile of compressed air. The LSOs and the 'meatball' are authentic, and it is a good place to begin the business of Marine Harrier II landings afloat.

got a vivid combat demonstration of the potential of V/STOL in 1982, when a handful of RAF ground-attack Harrier GR.3s and Royal Navy Harrier FRS.l fighters achieved remarkable results in the Falklands War.)

Although the US Marines hadn't been involved in the 1961 Kestrel trials, at least one senior officer became interested in the aeroplane's potential for supporting amphibious assaults. He directed two pilots to visit Farnborough in 1968 to investigate the next-generation Harrier. The result was an

order for 12 of what the Marines called the AV-8A, the first of which was delivered in 1971.

The aeroplane fits perfectly into the Marine's scheme of manoeuvre, and has been enthusiastically incorporated into the Corps. Over the years the Harrier has been gradually enhanced and improved so that the current version is far more powerful and capable than the fledgling which appeared more than twenty years ago. The Harrier II, as old as it is, seems destined to be a permanent part of the US Marine Corps family of

The tower is constructed of aluminium tube, and it wiggles a bit under the hot blast of the Harrier II's vectored thrust downwash.

aircraft for some considerable period of time, new generations following the old. The basic airframe and powerplant combination seems such a valid design that there is no real pressure for a replacement aircraft. Instead, the Harrier II continues to be refined and improved. As Marines like to say about many elements of their unique institution, the Harrier II exemplifies the spirit of *Semper Fi*.

Although the Harrier II could, as they say, operate out of a tennis court, about the only things it would be able to carry with it would be a few tennis balls. Vertical take-offs require a lightly loaded aircraft and consume lots of fuel. Instead, standard departures usually take the form of a short rolling take-off, sometimes with the aid of a ramp. With a maximum load of fuel and weapons aboard, the average short take-off (STO) distance is about 1,300 feet.

Short take-offs let the wing work, the result of which is vastly improved fuel economy and warload. For example, maximum aircraft weight for vertical take-off is about 18,000 lbs, with a warload of only 6,000 lbs. Short take-off gross weight goes up to 26,000 lbs. In this case the warload would probably be around 9,200 lbs of bombs, missiles and shells - sixteen Mk 82 500-pounders, or six Mk 83 1,000-pound bombs, plus a pair of Sidewinders, would be typical cargoes.

So it makes sense for the Marine ACE (Air Combat Element commander) to find a spot to park his Harrier IIs where they can get a running start before rotation. Roads are ideal for this, but a parking lot will do - just about any open space will suffice. But the Harrier II's downblast will tear up materials softer than concrete or steel plate, throwing them around and annoying the ground staff (not to mention feeding all that crud into the engine intakes).

This forward basing makes the Harrier II a success story with the Marines. Although Marines are officially an integrated part of the American armed forces and have the resources of the Air Force and Navy to call upon, historically they have been left on their own in some nasty brawls, with only Marine Corps assets available for help. Their mission today requires complete self-sufficiency, so they have their own dedicated air arm of Harrier IIs and Hornets.

A direct result of this force structure is that the ACE can put Harrier IIs ashore as soon as the amphibious assault waves secure about an eighth of a mile of roadway, turn them loose, and watch the show. With enough room for STO departures, the Harrier II can load up with fuel and bombs and then hit anything out to a range of about 300 miles from home, come back, and do it again. It may not be the fastest thing in the sky, but it will probably be the first thing overhead when a rifle company

VMA-211 'Wake Island Avengers' are the only Marine squadron to retain this star-and-meatball marking, a legacy from World War 2. Marines take their history seriously and teach it to every new recruit and officer candidate. Squadron histories are matters of great pride to most Marines, and small reminders of past glories are common in the Corps. (Note that some of the suction relief doors on the intakes are open on this jet.)

gets into heavy contact with the opposition. It can be dropping bombs on targets less than ten minutes after the unit gets the call if the squadron has aircraft sitting on the ground, engine off, on five-minute 'strip alert'.

The STOL capability means that it can be laagered on a road, a parking lot, a pasture, or any other piece of vacant real estate that is conveniently close to the 'forward edge of battle area', the FEBA. Once they have staked their claim Harrier IIs stand by on strip alert, waiting for the call from the forward air controller on the ground or in the air over the fight, on either one-hour, thirty-minute, or five-minute alert. 'If you're on five-minute strip alert, you have five minutes to be "gear in the well, gear up", *en route* to the target', said one of VMA-231's senior pilots.

All three types of alert involve having the aircraft loaded-up and fully fuelled, but the five-minute status has the pilot strapped into the cockpit, INS initialized, BIT (Built In Test) checks complete and the engine already run up and checked, then shut down. This is normally reserved for those occasions when the battle is raging over the next hill and the friendly services of the Harrier II squadron are a life-and-death issue. The APU (the little turbine auxiliary power unit) hums away, powering the aircraft's electrical systems and sipping lightly at the fuel supply. The pilots may sit in the cockpit for hours, monitoring the radios; then the call comes.

Beachhead mission

'Shank One, you're signalled to launch. Contact Raintight on 242.5'. You already have electrical power from the APU. Over by your left knee is the start panel: the START switch goes from OFF to START, the throttle goes from STOP to IDLE, and the rest is automatic. You monitor the gauges for over-temps; RPM comes up to where it belongs. Flap control is moved to STOL, and away you go.

'Paddles, Shank 11, taxy'. The Landing Signals Officer (LSO, usually called 'paddles') will position himself down the road where he can keep an eye on the proceedings. *'Roger',* he calls, *'you're cleared into position. Winds twenty port, ten knots. Check water and flaps. Cleared for take-off'.*

Off the deck and out of the traffic pattern, call the Direct Air Support Centre (DASC) - they are the folks who've invited you to the party. *'Raintight, Shank 11, airborne as fragged'.* 'As fragged' means as requested in the original call. For example, the DASC could have requested a flight of four with four Mk 83 bombs each, and that's what they get.

'Roger', they call back, *'Shank 11, contact Yankee One Victor on blue'.* 'Blue', according to the notes from your briefing, means 282.7, already programmed into the radios and just a button-push away. *'Yankee One Victor, Shank 11 with you, approaching ALPHA'.*

The AV-8B Harrier II does not have a radar (although they are working on putting the APR-65 in a version called the II Plus) . Instead, it carries in its prow the Hughes AN/ASB-19 Angle Rate Bombing System (ARBS), with a combined television camera/laser tracker-designator. The ARBS allows the pilot to designate an optically-acquired target through the use of a television system, which locks on to the contrast variations of the image. The computer uses the data to provide the pilot with steering and release cues in the HUD, and (when everything is working) these put the bombs on target. It is designed for low altitude deliveries, however, and in the Gulf (where high altitude missions were the norm, and contrast was low) the ARBS was not heavily used.

(Right) Mk 82 'pig iron' with tail fuses receive a final inspection from the squadron armourers. These 500lb bombs have been the American flyer's 'plain vanilla' attack ordnance for decades, which explains why these bombs bear manufacturing dates going back to 1972.

'Roger, ready to copy brief'. Normally there is a nine-line format for the brief, but it is often abbreviated to just the pertinent details. In the Gulf War, where close proximity to friendly troops was often not a factor, that part of the brief was omitted. And, instead of the usual forward air controller (FAC) on the ground, a back-seater in an F/A-18D normally performed the honours. This is because, if the Harrier II has a failing, it is poor eyesight from high altitude; it is a low-altitude system without a targeting FLIR (Forward Looking Infrared) or radar. So, with most strikers flying at

about 15,000 ft to avoid the shoulder-fired SAMs, the Harrier II needs the help of the Hornet with its radar and FLIR to find targets. The FAC-A (for 'airborne') will put the Harrier II flight in a holding pattern until he is ready to put them to work; then he will call.

'I've got seven T-55s in revetments just north of A0750. Find the north-south highway. Go north about a mile. You should see a large tower. One hundred metres east of that tower are the tanks. I'm coming in from the east; I will mark the target for you'.

The Hornet swoops down and puts a white phosphorus rocket on the target centre of mass to make sure you know where he wants your bombs delivered. The WP blossoms on the desert, a white cloud that's hard to miss.

In flights of two, diving from about 18,000 ft, you attack the tanks. Flight Lead will go first, then you drop, coming in about 30 seconds later on a different heading to make things a little tougher for the gunners on the ground. The FAC-A will try to spot your hits. Although you can't really see the tanks until you get down to about 11,000 ft, you can see the hits from your flight leader. The number two guy hears *'Dash Two, from Lead's hits go 150 metres south'*. The impacts are certainly easy enough to see, and you can adjust your aim point in the HUD on the way down.

The HUD will display the computed impact point as a bomb fall line and cursor; when they intersect your target, release the pig iron by depressing the red button on the left-top of the stick. You will feel them pop off, and if you haul back on the stick a bit the aircraft will begin heading back up again. You will pull about six Gs at the bottom - around 7,500 ft - but that's not too bad.

It's not a bad idea to reach over to the countermeasures panel and punch out some flares on the way up, since you are inside the range (although not by much) of the SA-7 Grail IR shoulder-fired SAMs which the enemy fields in lavish numbers. And the Harrier II is an IR missile magnet, with the engine exhaust right amidships rather than at the extreme stern, as with other fighters. A near-miss for any other fighter will probably take a wing off the Harrier II.

Well, that's about it for a Harrier II mission. You can go back and do it again, if there's anything left on the pylons or targets on the ground. You can even go down and use your guns, giving the SAM shooters an even better chance to return your favours. Eventually a female voice, computer-generated and highly annoying, will whine in your headset *'Bingo fuel!'* The pilots call her 'Bitching Betty', and she is always right; so it's time to go back to the barn. A nice hot shower, a change of uniform, and a cold beer at the O-club sound good? Forget it: even if you don't have to fly another mission, you camp out with the troops, sleep under the wing of the aeroplane, and instead of a cold beer you will get a canteen cup of the flavoured water that Marines call 'bug juice'.

Five-inch Zuni rockets put a lot of steel on a target. These, too, are old weapons, going back to the Vietnam era; but they seem to go off with the same old zest despite their antiquity.

555   555   556
556
STRAP
&
ATTACH

546L   547L

CRADLE
SUPPORT

(Inset) Mk 188 impact fuses installed on the 5 in. rockets. One of the things Marine armourers learn early in their careers is to avoid hitting these things with hammers.

Just prior to launch, a strap-on gun pack is inspected by one of the armourers. The gun is a 25mm five-barrelled GAU-12A 'Equalizer'. The ammunition (300 rounds) is fed from a separate pod on the starboard side of the aircraft.

514L 514L

NOSE ARMING
SOLENOID

TANK
FWD POST

MGP 331

These small Mk 76 practice bombs are painted blue (indicating that they are inert), but they actually contain an explosive cartridge which goes off on impact with the ground, marking the hit. Although the bomb is tiny, its flight characteristics mimic much larger free-fall weapons like the Mk 82.

This innocent-looking container is actually a 500 lb BDU-33 napalm bomb, a simple aluminium tank with 67 gallons of jellied gasoline inside; the two caps amidships are the fuses.

(Right) Nearly all Marine Harrier IIs have been painted in this fashionable and subtle grey-on-grey scheme selected by the official Corps 'interior decorator'. Actually, although camouflage

The tail navigation light and a pair of radar attack warning antennae adorn the stern of the Harrier II.

...chemes come and go, this colour ...eally does seem to make the ...ircraft almost invisible when ...ttacking ground targets. This ...etail shot shows the leading edge ...f the all-moving tailplane.

The wing features small fences that feed the airflow back to the control surfaces regardless of angle of attack. Wing structure is based on a triple-spar design; carbon fibre composites are used extensively to reduce weight (a saving of about 300 lbs), and they are spectacularly strong.

(Right, above) AN/ALE-39 chaff and flare dispensers mounted on the upper surface of the fuselage, aft of the wing. This configuration is unusual for American aircraft, but it is also found on the MiG-29 and Su-27 – a result of Afghanistan combat experience.

(Right) Yaw-control reaction air valve. Although the Harrier II can hover, that is about its only similarity to the helicopter; it flies in radically different ways, with different flight controls. At slow speeds engine bleed air is ducted out of these valves, providing control.

(Left) The accommodations in the Harrier II are quite similar to the Hornet, with programmable CRTs instead of analogue gauges. The HUD is of the wide-field variety and tends to dominate the view forward, but that was the idea anyway. The panel with all the buttons below the HUD is the up-front-controller, and is used to programme most of the systems aboard. Seating is moderately comfortable on the UPC/Stencel Type 10B ejection seat, with zero/zero capability.

(Above) The stick is embellished with switches for radios, weapons, trim, chaff and flare dispensers, and target selection and designation controls.

CIP AUT · BRT · TOO · VOL · I N 2 3 CLR · I/P · W 4 E 5 6 VOL · 7 S 8 9 – · ENT 0 . ON OFF · COMM I · COMM 2 · IFF TCN AWL WPN · WOF BCN ALT EM CON

HUD SYM BRT · STBY RET DPR · ALT BARO · NORM REJ 2 OFF · DAY NIGHT OFF · 0 RDR

ENGINE

(Left) Left console. The throttle has the usual HOTAS buttons and switches, and inboard is the nozzle control with its preset positions. The fuel panel is aft, and the stability augmentation system (SAS) controls are forward of the throttle/nozzle assembly.

(Above) Detail of the up-front-controller. This older AV-8B, predecessor to the night attack version, has only one multi-function display.

(Below) Teams of specialists care for the Harrier IIs, often working long hours, and never getting a ride. There are dozens of access panels on the aircraft, and when something needs inspection or repair all those fasteners have to be undone.

VMA 211

One of VMA-211's Harrier IIs
snoozes under the hot Arizona sun.

Capt. Dave 'Blade' Bonner climbs aboard one of VMA-231's aircraft at Cherry Point, North Carolina. The Harrier II is boarded from the right, unlike most American fighters, up an integral set of hand and foot holds.

(Right) Capt.Bonner flew 40 missions in the Gulf, is a Weapons and Tactics instructor and MAWTS-1 programme graduate. His squadron, VMA-231 'Aces', has been flying AV-8s operationally since 1975.

You can sit here for hours, and Capt.Bonner has done just that - up to three hours in the hot seat on five-minute 'strip alert'. Once you've fired up the engine, run through the BIT checks, aligned the INS and settled back to wait, the APU will keep the systems powered up, the INS fresh and the battery at full charge. When the call comes all you have got to do is switch the battery to ON, advance the throttle from STOP to IDLE, and you will be in business in less than a minute. You can be airborne in five, and dropping bombs shortly thereafter.

Directly astern of the pilot, generating huge quantities of noise along with a lot of power, is the Rolls-Royce F404 turbofan. This engine will pump out 21,500 lbs of thrust, vectored in just about any direction you want.

(Below) Here is the first stage of the compressor section. The blades are single-crystal forgings of tremendous strength and size.

AN/ALQ-64 electronic countermeasures pod on the number four centreline station.

A lightly loaded Night Attack Harrier II returns to its nest at Cherry Point, North Carolina, after an hour-long sortie over the local training area.

Light on fuel and weapons, a Harrier II from VMA-542 'Flying Tigers' descends to the pad at Cherry Point. Vertical landings involve an approach to a hover about 30 ft above the deck, then a smooth, deliberate descent all the way down. During the descent the pilot will be easing the throttle aft, maintaining his position with stick and rudder pedals. It isn't like landing a helicopter! At slow forward air speeds, without the effect of ram air, the engine suction opens eight inlet doors in the intake cowling.

(Left) Harrier II drivers wear a wide variety of patches, some sacred, others quite profane; this is a common one.

(Left, below) VMA-211 'Wake Island Avengers' wear this patch. On the day following the Pearl Harbor attack the squadron (then called VMF-211) attempted to defend the small Pacific island against a Japanese attack with their 12 brand new Grumman F3F Wildcats. Seven were shot up on the ground during the first enemy strike; the other five fought off numerous Japanese attacks, shooting down many enemy aircraft during the next two weeks. Gradually the remaining Wildcats were destroyed, and the garrison was forced to surrender on 22 December 1941.

(Below) VMA-311 'Tomcats' also prowl around out of Yuma, and they too trace a proud squadron history back to World War 2. The unit flew F4U Corsairs for much of the conflict, and pioneered the use of the aircraft in close air support; VMF-311's dive-bombing exploits against Japanese targets in the Western Pacific campaign are legendary. The unit performed the first US jet combat mission in Korea on 12 December 1950, with F9F Panthers; and flew over 50,000 combat sorties over Vietnam.

The 'Tomcats' again, this time on the flight deck of one of the specialised assault carriers which the US Navy operates just to keep the Marines happy – in this case USS *Belleau Wood* (LHA-4) – somewhere out in the Pacific Ocean during *Operation Tandem Thrust*.

You can sit around waiting for the 'squids' to get their act together for hours out here, but sooner or later they will probably decide it's launch time. Then they point the ship into the wind, and the LSO invites you to leave. While it may not be as dramatic as a 'cat shot' off one of the 'big boats' in an F-14, this little aircraft will nevertheless put what the Marines call 'serious hurt' on the enemy ashore in a great hurry.

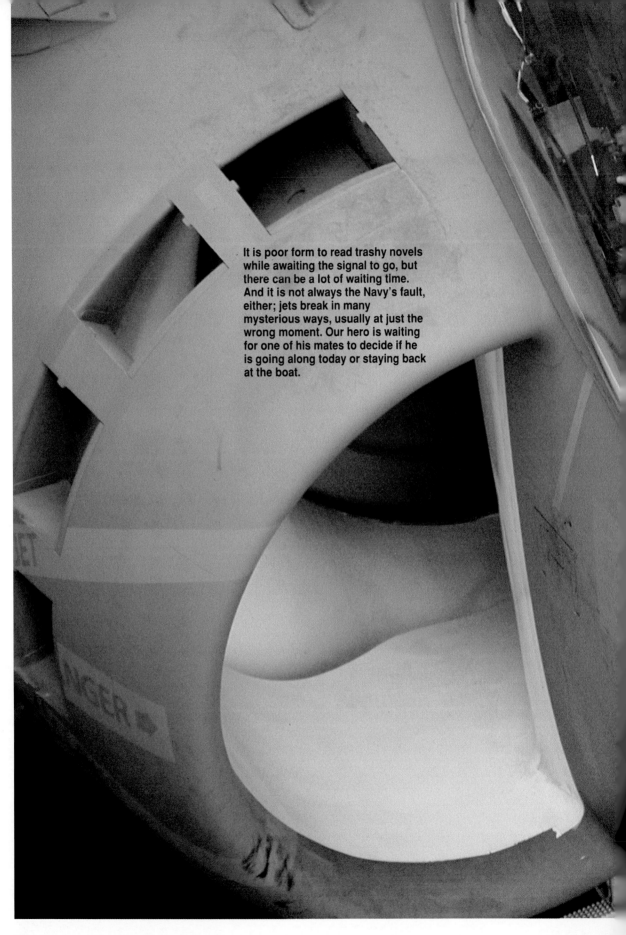

It is poor form to read trashy novels while awaiting the signal to go, but there can be a lot of waiting tIme. And it is not always the Navy's fault, either; jets break in many mysterious ways, usually at just the wrong moment. Our hero is waiting for one of his mates to decide if he is going along today or staying back at the boat.

Carrier departures are done with the guidance of the Launch Officer (LO). At his signal, this Harrier II taxies into position at the 300 ft mark on the deck centreline. The LO will signal run-up; that means to advance the throttle to 60 per cent and drop the nozzles to the STO (short take-off) stop, usually at 50 degrees. The LO checks the flap position and trim setting, then signals to the pilot 'okay!' The nozzles then come back to 10 per cent.

Well, finally! Canopy down and locked, aligned with the centreline, nozzle stop set and water switch set. Then take your signals from the LSO . . .

The signal is 'Hold position!'

'Okay!', signals one of the LSO team.

Ready for the 'active', the number two aircraft in the section waits for the flight lead to get out of the way. Once directed to the spot the pilot will salute the LO, indicating that he is airworthy. Then the Launch Officer will watch the pitch and roll of the deck, waiting for an opportune moment to genuflect, touch the deck, and order the Harrier II off the ship. The pilot advances the throttle to MAX and pops the brakes; as the AV-8B crosses the STO line (a yellow band across the deck) he slams the nozzle control back to the pre-set stop, usually 50 degrees, and roars off into the sunrise.

93

Recoveries are made by sliding the aircraft in from the ship's port side, then settling down firmly amidships, beside the island.

A view from the 'Vulture's Roost', the nam
given to the upper deck where spectators
gather to watch the proceedings. The
small, thin wing root extension is
sometimes referred to as the 'tin wing'. Tl
canopy of the Harrier II has just about the
heaviest charge of embedded explosive c
any combat aircraft anywhere. That
squiggly line over the pilot's head will tak
the Plexiglas canopy apart a few
milliseconds after the pilot tugs on the
ejection seat ring, and a few more
milliseconds before the rocket motors
under the seat propel the resident aviato
skyward, *sans* aeroplane.